Looking at Energy

Don Radford

Batsford Academic and Educational Ltd London

© Don Radford 1984
First published 1984

All rights reserved. No part of this publication
may be reproduced, in any form or by any means,
without permission from the Publisher

Typeset by Tek·Art Ltd, West Wickham, Kent
and printed in Great Britain by
R J Acford
Chichester, Sussex
for the publishers
Batsford Academic and Educational Ltd,
an imprint of B.T. Batsford Ltd,
4 Fitzhardinge Street
London W1H 0AH

ISBN 0 7134 3486 4

Contents

- 4 Acknowledgment
- 5 Hymn to the Sun
- 6 Looking at the "Go" of Things
- 8 Forms of Energy 1
- 10 Forms of Energy 2
- 12 Muscle Power Rules, OK?
- 14 Matches, Rockets and Explosives
- 16 Steam Engines
- 18 Internal Combustion Engines
- 20 Turbines and Jet Engines
- 22 Chemico-Electrical Energy
- 24 New Chemicals for Old
- 26 Solar Energy
- 28 Water Power
- 30 Windmills and Sailing Ships
- 32 The Force that Drives the Sun
- 34 Gravitational or Down-Hill Energy
- 36 Other Sources of Energy
- 38 Storing Mechanical Energy
- 40 Storing Electrical Energy
- 42 The Changing Faces of Energy
- 44 The "Go" in our Lives
- 46 Glossary
- 47 Book List
- 48 Index

Acknowledgment

The Author and Publishers thank the following for their kind permission to reproduce copyright illustrations: Alastair T. Arnott, for the drawing on page 39; Austin Rover, for the photograph on page 18; David Baker, page 15 (left); British Airways, page 21; BBC Hulton Picture Library, pages 7 (top), 10, 17 (left), 19, 35, 38 (left); BBC Photographs, page 9 (bottom); Central Electricity Generating Board, page 20; The Electricity Council, page 41 (bottom); Farmers Weekly, page 45 (top left); Ford Aerospace and Communications Corporation, page 27; Pat Hodgson, pages 6, 11 (bottom), 14, 23, 25, 38 (right), 42, 45 (top and bottom right); Lucas, page 41 (top); National Coal Board, page 45 (bottom left); North of Scotland Hydro-Electric Board, page 29; Science Museum, London, page 17 (right and bottom); Science Photo Library, pages 7 (bottom), 32; Shell, page 24; Jamie Taylor, Edinburgh University, page 36; John Topham Picture Library, pages 9 (top), 12, 15 (right), 30; UK Atomic Energy Authority, page 11 (top); World Health Organization, page 34. The pictures were researched by Pat Hodgson. The diagrams were drawn by Rudolph Britto.

Hymn to the Sun

Thou appearest beautifully on heaven's horizon
O living Aten, the beginning of life.
When thou art risen in the east
Thou fillest every land with thy beauty.
When thou settest in the west
The land is in darkness, as in death.
Darkness is a shroud, and the earth is still,
For he who made it, rests below the horizon.
At daybreak, when thou hast risen
When thou shinest as Aten by day,
Thou drivest away darkness and givest us thy rays.
Then all beasts are content with their pasture;
Trees and plants flourish,
Birds fly from their nests,
Their wings outstretched in praise to thy spirit.
All beasts and all that fly
Live when thou hast risen for them.
Ships sail north and south,
For every way is lighted by thee.
The fish in the river dart before thy face,
Thy rays are even in the midst of the great green sea.
Thy rays feed every meadow.
When thou risest, all live, all grow for thee.
Thou makest the seasons so all should grow;
The winter to cool us,
The summer heat that we might taste thee.
Thou didst create the earth and every thing therein,
All men, cattle and wild beasts,
Whatever goes on its feet,
And what is on high flying with wings,
Thou hast made them.

(by the Egyptian Pharaoh, Akhenaten, circa 1370 B.C.)

Looking at the "Go" of Things

What did you have for breakfast this morning? Was it Corn Flakes? One well-known brand says on the side of the packet: "A good, nourishing breakfast is the most important meal of the day, and the ENERGY and vitamins it can provide will give you and your family the best possible start to each day."

What is ENERGY? Are you energetic? Over a hundred years ago a great scientist, James Clerk Maxwell (1831-79) described energy as being "the go of things". Everything that lives, moves or changes in any way, depends upon energy. Without it we would not be able to hear the songs of birds, enjoy our favourite bangers and chips, or see the colours of the rainbow. Without it we would not be alive. Everything we do requires energy; even sitting still reading, or whistling a tune, or even sleeping. All need energy. Without energy the whole of the Universe would be silent, dark and still.

Energy is the key to the understanding of everything around us. Yet it is only just over a hundred years since man began to recognize its importance. The discovery of fire many thousands of years ago and the use of simple machines to harness muscle power started man on his long journey from the bow and arrow to the rocket, his space craft to the stars. As man developed and learnt more of the world around him, so did he use more and more energy. At first, he had only his own muscles. Later, he employed animals to help him. Later still, he learnt to use the forces of nature. Water and wind-powered mills were built. Today, we harness the powers of the lightning bolt for our electricity supplies and attempt to control the forces of the sun itself.

We all use energy in our homes, schools and in travelling here and there. Industry uses even more energy, making things and transporting

them. Each of us uses the amount of energy that could be produced by 100 slaves working twelve hours a day. Today, in the Western World, we rely on machines to be our slaves and save us from hard, back-aching work. Machines do all the work that otherwise would have to be done by hand. But our modern slaves need feeding. Their food is energy.

Energy is the "go" of things. Often we take this "go" for granted, because energy is hidden. It wears half a dozen different hats and a thousand disguises. So, we must learn to probe and keep asking Clerk Maxwell's question, "What makes it go?".

Look again at the three illustrations on these pages. See if you can say how and where energy is involved in each object photographed. Remember that, if you can see or hear something, or if it moves, or changes, then energy is at work.

Forms of Energy 1

Mechanical Energy
This is probably the most obvious form of energy. A stone breaking a window or a speeding car colliding with another has energy. The faster the stone or car moves, the more is the damage. Likewise, the heavier the object, the more energy it has, if it moves at the same velocity.

Anything that moves has Kinetic Energy. Even the particles of a stone lying still on the ground have Kinetic Energy, because, although the stone may be still, the particles are vibrating. The hotter the stone, the more the particles vibrate and so the stone has more Kinetic Energy. In an engine, various parts may rotate or move to and fro. These parts have Kinetic Energy. We give the name Mechanical Energy to this form of energy. It is only Kinetic Energy wearing another hat. Whenever you see something move, there you see Kinetic Energy in action, though it may be called Mechanical Energy. The Sun, Moon and stars all move; so do cars, trains and aeroplanes; all have Kinetic Energy.

Electrical Energy
In our homes, switches control lights, electrical heaters, radios and electric motors. We can see or feel or hear something happening when we switch them on. In other words, we are aware of the effects when Electrical Energy is changed into another form of energy. In the case of an electric mixer, Electrical Energy is changed into Mechanical Energy. Electricity makes the motor go. With an electric fire, Electrical Energy is turned into a form we can feel. We call it heat. When heat falls upon an object, it causes the particles in the object to vibrate more vigorously. The object gets hotter. If the object gets too hot, then it might melt or boil or a chemical reaction may take place.

Electrical Energy depends upon the forces between electrical particles. Small negative charges of electricity called electrons can move along a piece of copper wire. We cannot see these electrical particles but can see or detect the effect they have when they move. It is the effects produced by moving electrons that operate our electrical equipment.

Electro-Magnetic Radiation
Radio Waves, Heat, Light and X-rays are all members of the same family – Electro-Magnetic Radiation. Electro-Magnetic Radiations consist of waves of energy. These waves travel through space at 3×10^8 metres per second. The distance

between successive peaks or troughs is called a wave-length (λ). Wave-lengths vary according to the kind of EMR. Radio waves have lengths that can be thousands of metres long. BBC Radio 4 is transmitted on a wave-length of 1500 metres, using long aerials strung between tall steel pylons. On the other hand, gamma (γ) radiations have very short wave-lengths and are produced inside the nuclei of atoms. Atomic nuclei are incredibly small, being about 10^{-14} metres. This distance is beyond understanding. 10^{-14} of the distance between the Earth and the Sun is about 1.5 millimetres.

The important thing to note about EMR is that this is the way we receive energy from the sun, or light from the page you are reading at present. Without EMR, Earth would be cold and dark and you would not be able to see.

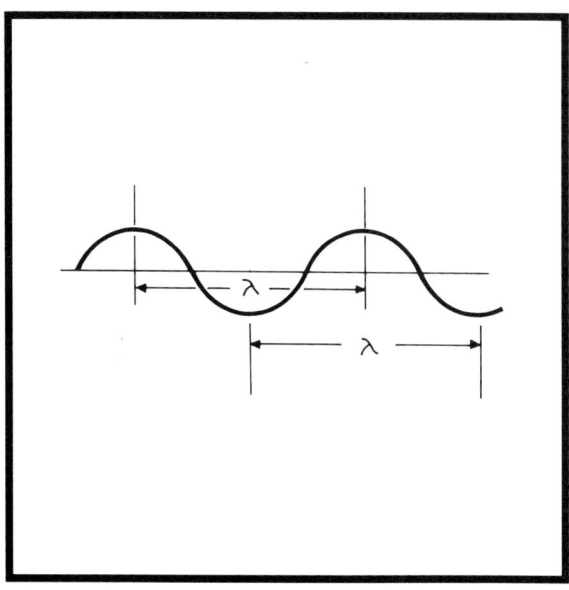

◁ Wave-length of electro-magnetic waves.

Forms of Energy 2

Chemical Energy
Whenever you see a firework display, burning gas, or a precipitate forming in a test tube in the laboratory, there you see the effects of the release of Chemical Energy. Your school bus or family car is powered by controlled explosions of fuel mixed with air inside the cylinders of the engine. The Chemical Energy released is converted into Mechanical Energy and this is transmitted to the wheels through a train of gears. Switch on a torch and a bulb lights. Chemical Energy stored in the battery has been released. It has then been converted into Electrical Energy, which has heated the thin wire in the bulb so hot that light is given off.

Even life itself in all its forms depends upon Chemical Energy. Chemical Energy is involved in the forces that bind atoms together to form molecules.

Nuclear Energy
Nearly everyone has a picture in his mind of Nuclear Energy. Maybe they think of an atom bomb or a nuclear-powered submarine or a nuclear power station, like Calder Hall, but few give a thought to the Sun. It is powered by Nuclear Energy. Even fewer people wonder why the interior of the Earth is very hot. Its heat is due to Nuclear Energy.

Nuclear Energy is involved in the forces that bind together particles, such as electrons, protons and neutrons, to form atoms. Nuclear Energy is the ultimate source of nearly all the energy we use. Remember, the Sun is a giant nuclear power station and we receive all its bounties by EMR (Electro-Magnetic Radiation) (see page 8).

Gravitational Energy
We have mentioned briefly forces that bind together subatomic particles to form atoms, and forces that join atoms to form molecules. Gravitation is the force that binds worlds together. When we lift a weight, we have to use Chemical Energy in our muscles to overcome the attraction of the Earth and the weight for each other. Not only does the Earth attract the weight, but the weight attracts the Earth. When we have lifted the weight, it has a certain amount of stored energy. This is called Potential Energy.

Let the weight go and gravity will pull it down; Potential Energy is converted into Mechanical Energy, the energy of movement. The Moon revolves round the Earth and is kept in its orbit by gravity. The Earth revolves round the Sun, and the Sun around some centre in the middle of our Galaxy. All are kept in their paths by gravity, the force first investigated scientifically by Sir Isaac Newton.

But what connection has Big Ben with gravity? Find out what makes it work. How is it powered? What makes it go?

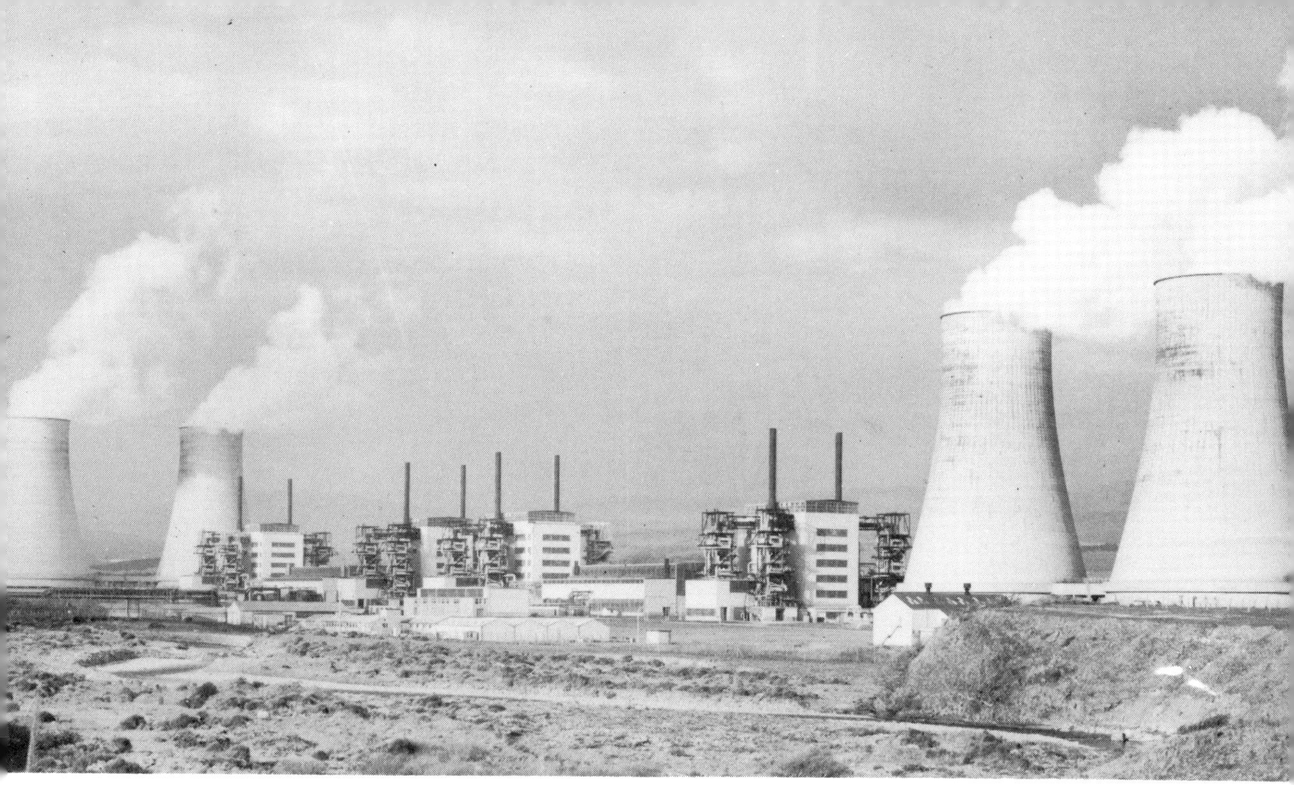

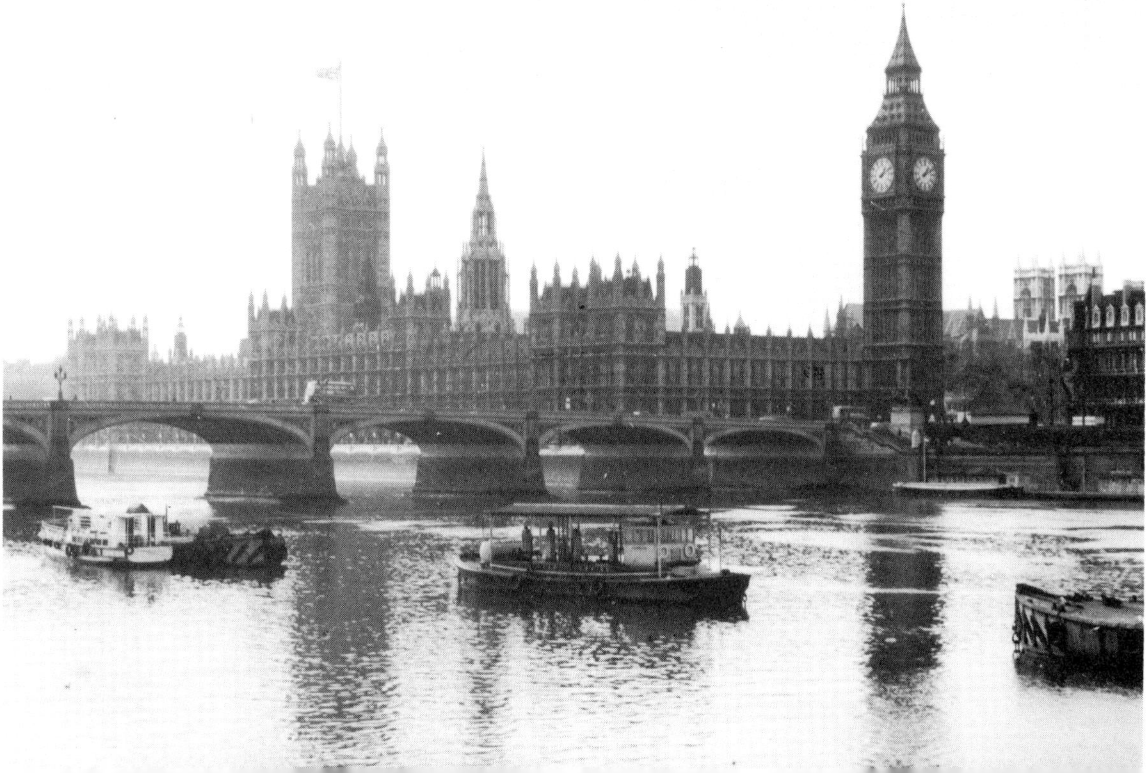

Muscle Power Rules, OK?

All living things need energy to do such things as growing, forming more individuals and replacing worn-out parts. Some living things need energy to move and keep warm, and others need it to make food. In a series of chemical reactions green plants use energy from the Sun, and carbon dioxide (CO_2), water (H_2O) and mineral salts to build new chemicals. These new chemicals have more Chemical Energy stored within their structure than was in the original substances. The extra energy comes from the Sun. The plants use these new chemicals, such as sugars, starch and cellulose, for growing and reproduction.

Animals are able to liberate energy by reacting food with chemicals derived from the oxygen in the air. That is why we need to breathe. We take in oxygen and send it round the body in our blood, to where it is needed. There it reacts with simple food chemicals, giving off energy. In the muscles the released Chemical Energy enables us to move about, work and play games. Heat is also produced and this keeps us warm.

Before the steam engine, man was entirely dependent on natural sources of energy. The first source to be used was muscle power. In the beginning, man had to use his own muscles, but sometime about 10,000 years ago he began to train animals to carry and pull things. Muscle power reigned supreme until the beginning of the Christian era. Then, natural sources of energy such as water and wind power helped man – until steam took over in the eighteenth century. Nowadays, man uses his muscles mainly for movement and for sport.

As an energy machine, man is neither prolific nor efficient. The average adult can generate about 0.1 Horse Power over a period of time; or,

put another way, he could just about keep a 75 watt electric light bulb alight, if he were to operate an electricity generator. It would take 40 slaves pedalling like mad to produce enough electricity to operate a 3 KW electric kettle or electric fire. It is just possible for a trained athlete in a super-lightweight aircraft to fly, using his own muscle power – but only just. In Man's history he made huge leaps forward when he learned to harness energies other than his own. First of all, he learned to use animal muscle power to carry him and pull his plough and chariot. Today in the Western World the horse is mainly used for sport – racing, hunting and jumping.

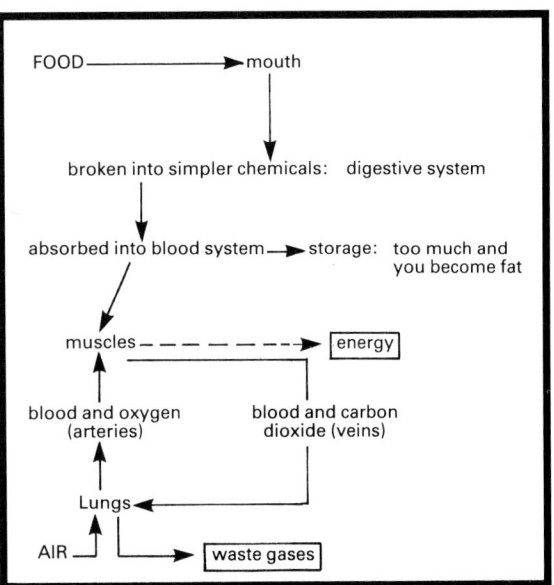

Matches, Rockets and Explosives

What is the connection between lighting a match, sending a rocket to the Moon and blowing up the Houses of Parliament with gunpowder? When we think about it, we find that there is more than one connection.

1. They are all examples of Chemical Energy in action.
2. In each case the chemical reaction produces a flame.
3. Nearly all burning things need oxygen in order to burn. In these examples, air is not needed to supply the oxygen. The head of a match or an explosive like gunpowder relies on oxygen in a suitable compound. A rocket motor often uses liquid oxygen. All this means that a match head or gunpowder or rocket fuel will all burn in a vacuum, or even under water, because they carry their own oxygen supply.

Let us look at the uses of these three reactions:

1. When a match is struck, the chemical reaction produces a flame that sets alight the matchstick.
2. With an explosive, the chemical reaction produces a large volume of very hot gas in a very short period of time. The struggling of this volume of gas to find space for itself produces the explosion; the quicker the reaction, the bigger the bang, and the more it can push out of its way.
3. The reaction that takes place in a rocket motor is like a long and controlled explosion. Large volumes of very hot gases are squirted rearwards. This drives the rocket on. Using propellers or jet engines, aeroplanes also move by squirting large amounts of air behind them.

In each of the three examples, energy is given out in the chemical reactions. The products of the reactions contain less energy within their structures than the original substances contained. Such reactions are called Exothermic Reactions.

15

Steam Engines

Few subjects rouse so much enthusiasm as the sight of a steam rail locomotive. The sight and sound of clouds of steam escaping and the size and colour of the engine all give an impression of power — power ready to be let loose at the movement of a lever. In Britain today, the steam locomotive is a thing of the past, seen only in museums and on private railway lines. Although still widely used in the Third World as steam locomotives and power sources for industry, steam engines are slowly being replaced by other sources of energy, such as electric motors and internal combustion engines. Yet it was the steam engine that helped to bring about the Industrial Revolution. It changed Britain from a poor, agriculture-based power into a country whose resources to make things were greater than the rest of the world's put together.

The Age of the Steam Engine began in 1705. It started in a most unlikely place — not where industries were gathered together, but in rural Devon. Thomas Newcomen, a blacksmith of Dartmouth, made the world's first piston-driven steam engine. It was designed to operate pumps to drain the tin mines of Cornwall. Wherever miners dug for coal and minerals, water seeping into the mines was always a threat. The deeper the miners dug, the greater was the chance that the mine would fill with water. Primitive pumps were used, but by the eighteenth century many mines were reaching the stage when they would have to be closed down. Water seeped in faster than it could be pumped out using muscle, wind and water power. Thomas Newcomen's engines enabled mines and miners to prosper and produce in plenty the raw materials for the manufacture of iron and steel.

The second half of the eighteenth century saw the development of steam engines by men such as Watt and Boulton. Then, in the nineteenth century, the steam engine really came into its own. On land and sea it powered railways, traction engines and ships, and provided the energy to work countless factories. The days of the horse-powered stage coach and the graceful, white-sailed tea clippers came to an end. For a while, King Coal and his henchman, Snorting Steam, ruled supreme, until they too were edged off their thrones by the thunder and lightning of the Internal Combustion Engine and the Electric Motor. The age of the motor car and aeroplane had arrived.

The first steam engines were devised by rule of thumb. Scientific reasoning played little part in their design. It wasn't until James Watt started to improve the steam engine in the latter half of the eighteenth century that science was brought in and the efficiency of the engine was raised from a miserable $\frac{1}{2}$% (that meant that, for every 100 tons of coal used in heating the boiler, only half a ton was used in producing useful work, and the rest was lost heating the surroundings) to about 3%. Today, the steam electricity generating stations of the C.E.G.B. operate with an efficiency in excess of 30%. Science has made possible the production of cheap power. Now, one ton of coal in a modern steam engine can do the work of sixty tons of coal used in a Newcomen engine. Such is the progress due to the applications of scientific knowledge.

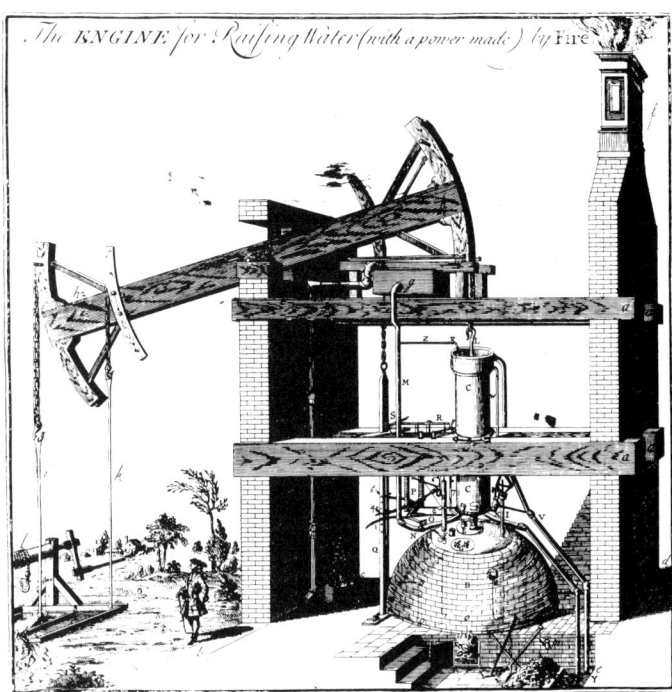

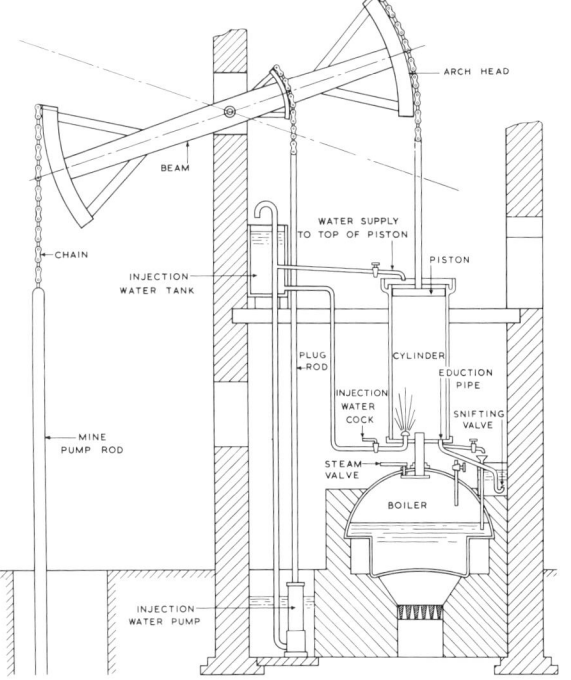

A sectional diagram of a Newcomen engine, ▷ showing how it worked. Briefly, steam was fed into the cylinder via the steam valve. This caused the piston to rise, aided by the weight of the mine pump rod. (The pressure of the steam alone was not sufficient to force the piston upwards.) When water was injected into the cylinder from the injection water cock, the steam condensed, causing a vacuum, and the air pressure outside drove the piston down; this in turn lifted the mine pump rod and operated the pump.

Internal Combustion Engines

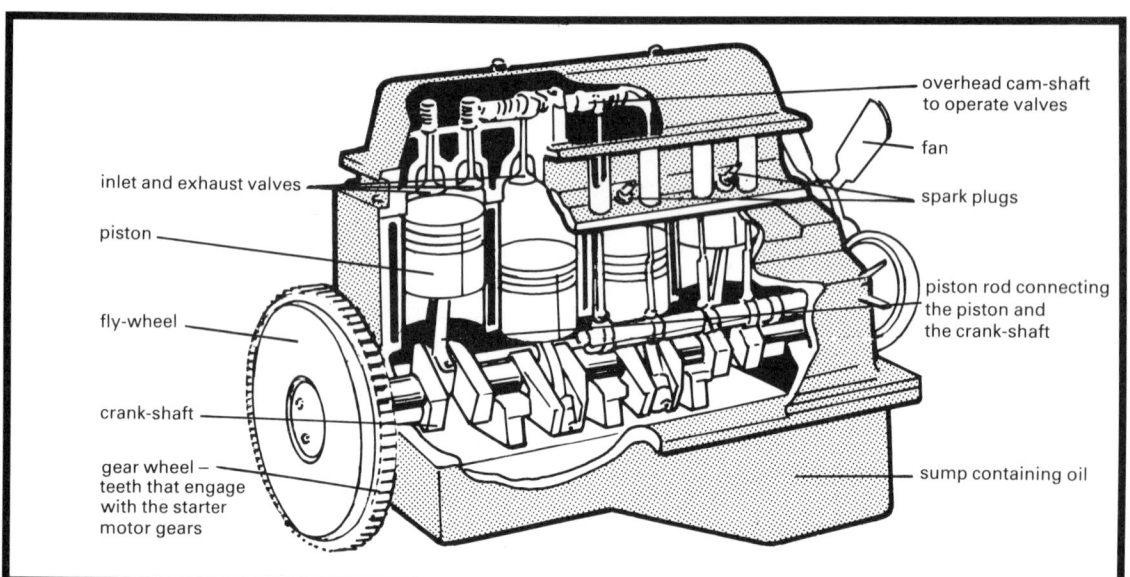

△ Car engine.

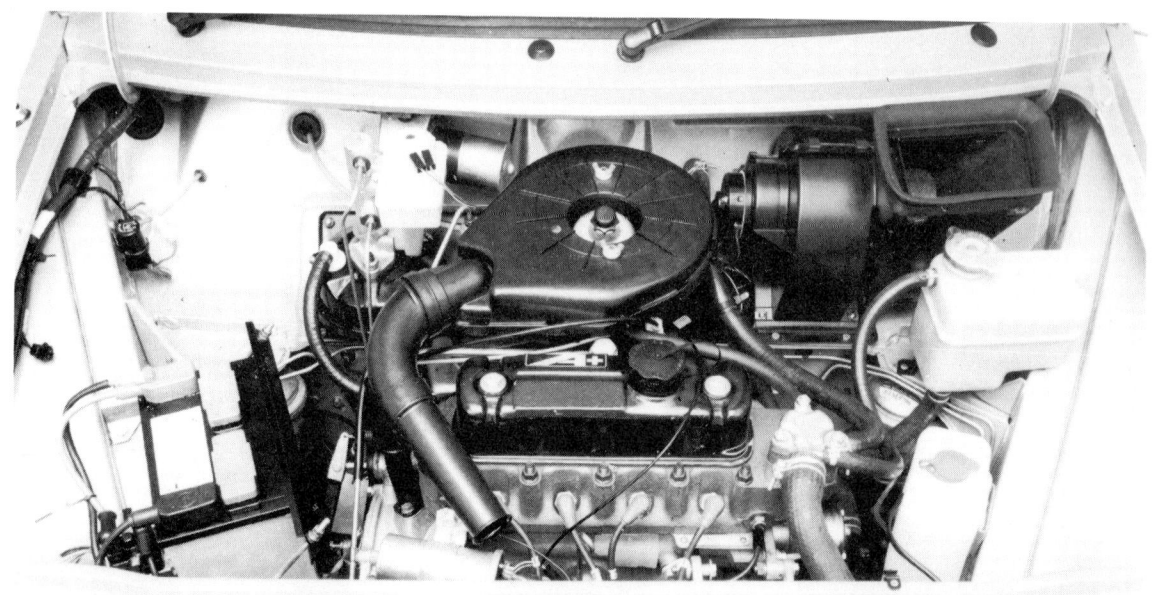

The steam engine, powering factories, ships and railways, had its disadvantages. It was heavy, relatively slow-moving and needed a boiler and furnace. During the last forty years of the nineteenth century the Internal Combustion Engine was developed. With time, its power increased and its weight per horse power decreased. This reduction in weight made possible first of all the motor car and, later in the twentieth century, aeroplanes. The Wright brothers' success in being the first men to fly and control a heavier-than-air aircraft was largely due to their being able to design and construct a lightweight internal combustion engine.

In many ways, the internal combustion engine used in our cars resembles the steam engine. It has cylinders, pistons and a crankshaft. But it differs in one important way. There are no boilers and no furnaces. Fuel is burnt inside the cylinders, not in a furnace. That is why gas, petrol and diesel engines are called *internal* combustion engines.

Both gas and petrol engines must have spark plugs and a source of high voltage electricity to fire the mixture of air and fuel. The high voltage is generated by the ignition coil and is distributed to the right spark plug at the right time by the distributor. Sadly, the ignition system is one of the principal causes of car breakdowns.

In 1892 Rudolf Diesel in Germany patented an engine that did not require spark plugs. It works on principles that are known to anyone who has pumped up a bicycle tyre. The pump gets hot. This is because the air has been compressed. Compressed air gets hot, expanded air gets cold. These two facts are behind the working of rain clouds, refrigerators, high-speed aircraft and Diesel engines. In the cylinders of a Diesel engine the air-fuel mixture is compressed to pressures of up to 300 lbs per square inch. This sudden increase in pressure raises the temperature of the gas so high that it catches fire. Aeromodellers have used small internal combustion engines working on this principle for the past forty years.

Any aircraft that flies above the speed of sound has heat problems. At high speed, air is compressed along the leading edges of the main wings and tailplane and causes the structure to reach temperatures at which ordinary metal alloys lose their strength. All these are examples of getting 2½ gallons into a pintpot!

In the 1920s the fastest aeroplanes were those racing for the Schneider Trophy. It was through the ability of a firm like Rolls Royce to produce a light but powerful engine and of Mr R.J. Mitchell who designed the Supermarine sea planes, that Britain won the coveted trophy outright, flying planes like the one in the photograph. Later, Mitchell and Rolls Royce combined to produce the Spitfire, one of the most famous planes of the Second World War.

Turbines and Jet Engines

For about 800 years man has been able to obtain energy from the winds by using windmills. A steam turbine obtains energy from rushing steam, as it is released from high pressure. A jet engine obtains energy from the hot gases produced by burning fuel. The gases expand as they pass through the gas turbine. By slowing down a stream of steam, air or gases, we can obtain useful Mechanical Energy.

The difference between a steam turbine and a gas turbine or jet engine is that a steam turbine needs a boiler and furnace to produce steam, whereas a jet engine or gas turbine burns fuel within itself. The difference is the same as that between steam engines and internal combustion engines.

Steam turbines are used to provide energy to drive ships and electrical generators. Jet engines and gas turbines work on the same principles to produce energy, but make use of the energy in different ways.

Jet engines burn fuel and use the energy to suck in air, heat it and blow it out again so that it pushes an aeroplane, such as Concorde, in rather the same way as the gases squirted out from a rocket thrust a space craft forwards.

Gas turbines also burn fuel and use some of the energy produced to suck in air. But most of the energy is used to drive a turbine in a similar way to a steam turbine. The energy from the gas turbine can be used to drive propellers in aircraft and fast-moving warships, and to turn electrical generators.

The great advantage of gas turbines is that they can be started from cold and, in a few minutes, be producing full power to drive a warship or an electrical generator. A steam turbine using a furnace and boiler could take eight hours before producing full power.

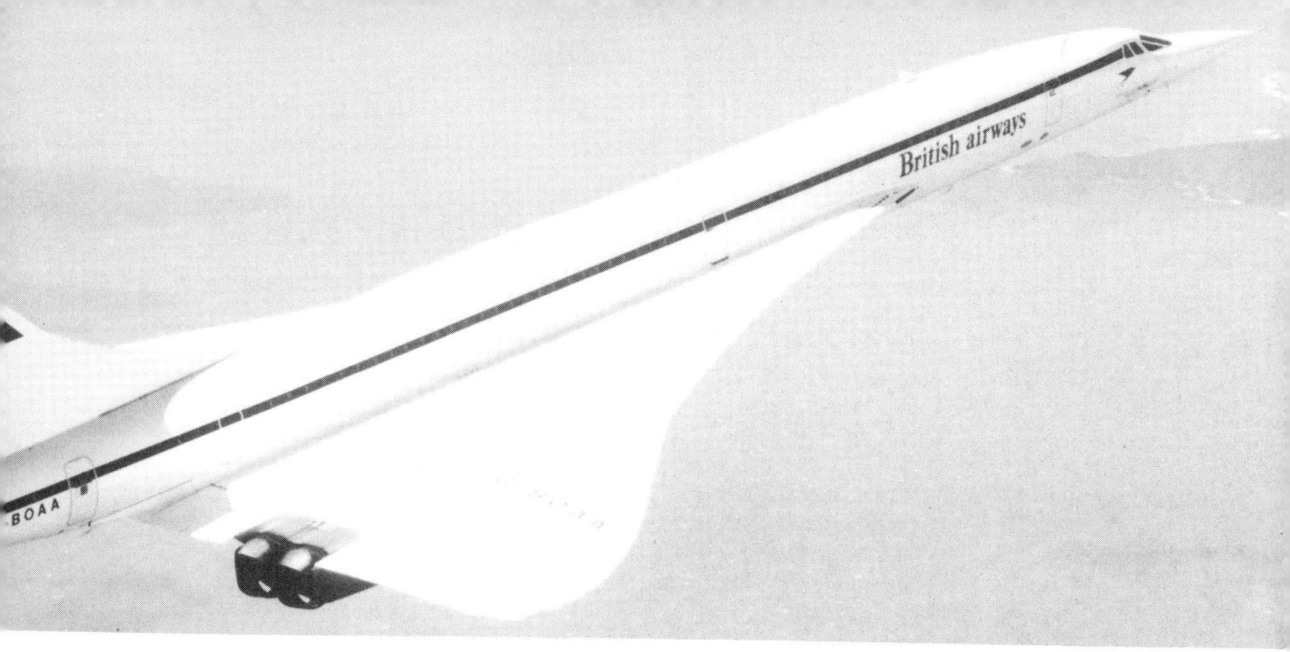

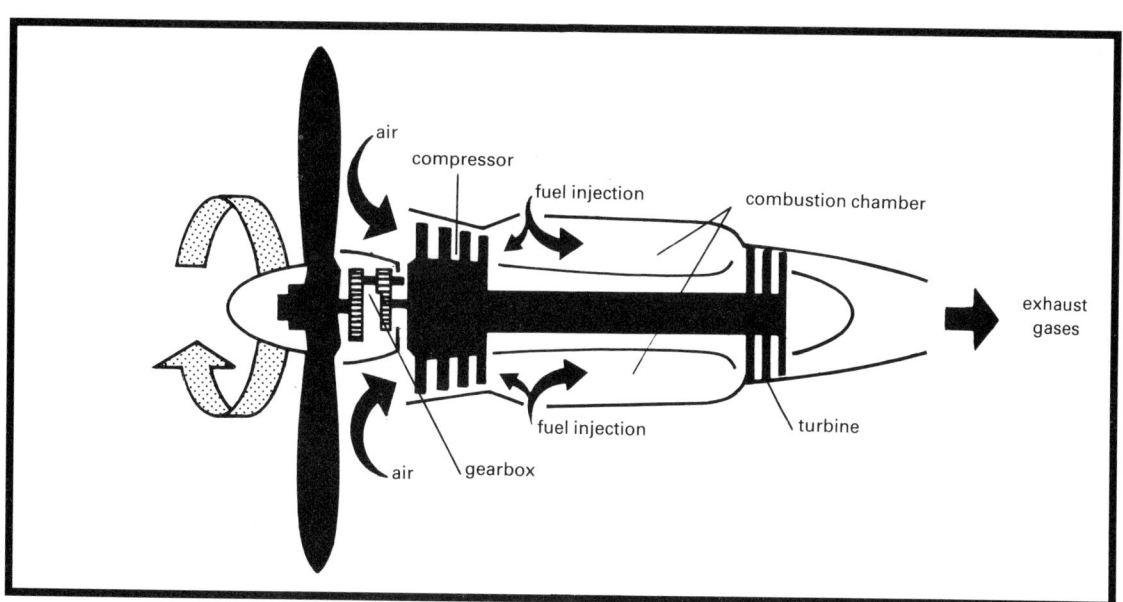

1 This is one of the steam turbines at the Central Electricity Generating Board's Ironbridge plant. It is capable of producing two thirds of a million Horse Power, or 500 million Watts of electricity. This amount of energy would be sufficient to heat half a million single bar electric fires, or run ten thousand family cars. For an idea of the size of the turbine, look for the mechanic working on the front end.

△ Turbo-prop aero-engine.

Chemico-Electrical Energy

What do you need to make your transistor radio go? What do you have to replace when it runs down? What powers your torch or your calculator or digital watch? The answer is the same in each case – a battery. A battery is a device that changes Chemical Energy into usable electricity. It has two connections which need to be joined to whatever you want to "go". One connection is called the positive electrode and is marked on many batteries by a + sign. The other connection is called the negative electrode and is marked by a − sign. When the + and − electrodes are connected by a piece of copper wire, small electrical charges called electrons move through the wire from the negative electrode to the positive electrode. Meanwhile, inside the battery, chemical reactions take place at the two electrodes. In one reaction, electrons are produced and the reaction will only continue if they are removed. The other reaction will only proceed if electrons are available. Thus the two reactions depend on each other and one won't go without the other. It is a case of "you rub my back and I'll rub yours!"

All batteries have three parts: two dissimilar substances, such as zinc and copper, or zinc and a carbon, manganese dioxide mixture; and something called an electrolyte. It is in the electrolyte-electrode junction that the reactions take place. Once upon a time, a battery meant a "dry" cell and all dry cells were the same, with electrodes of (zinc) and (carbon and manganese dioxide) and an electrolyte of a paste of (starch, water and ammonium chloride). Today, there are nearly two dozen different types of batteries, most of them containing very unusual chemicals. Lithium, the most reactive of all metals, is the negative electrode in many of them.

If we have a fuel, such as methanol, we could burn it in an internal combustion engine or steam engine to drive a dynamo to produce electricity. The Chemical Energy would be converted into Heat Energy, then into Mechanical Energy and then into electricity. The theoretical efficiency of such a chain of changes is less than 40% and the practical efficiency would be even lower. The main loss occurs in the conversion of Heat Energy into Mechanical Energy. If we could go, in one stage, from the Chemical Energy (in methanol) to electricity, then much of the losses would be eliminated.

It is possible to do this in a fuel cell. Unfortunately, the practical difficulties in using common fuels such as coal, natural gas and petroleum have not been overcome so far. Some space craft have been fitted with fuel cells, but these use gaseous oxygen and hydrogen. Attempts have been made to power electrical cars with fuel cells, but the snag has been that these cells have had to be too large.

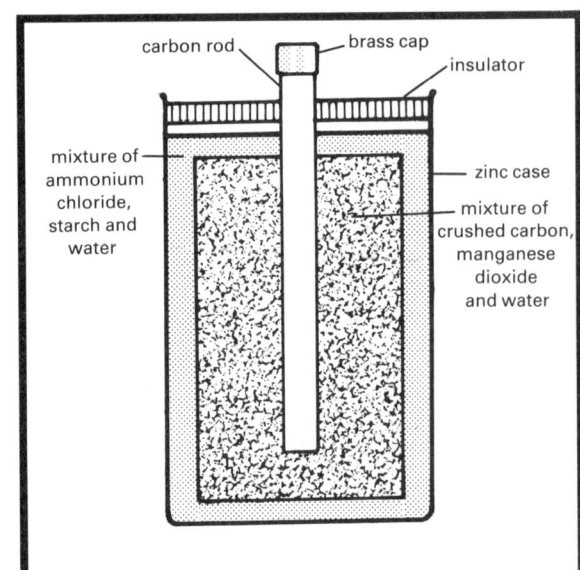

A dry cell.

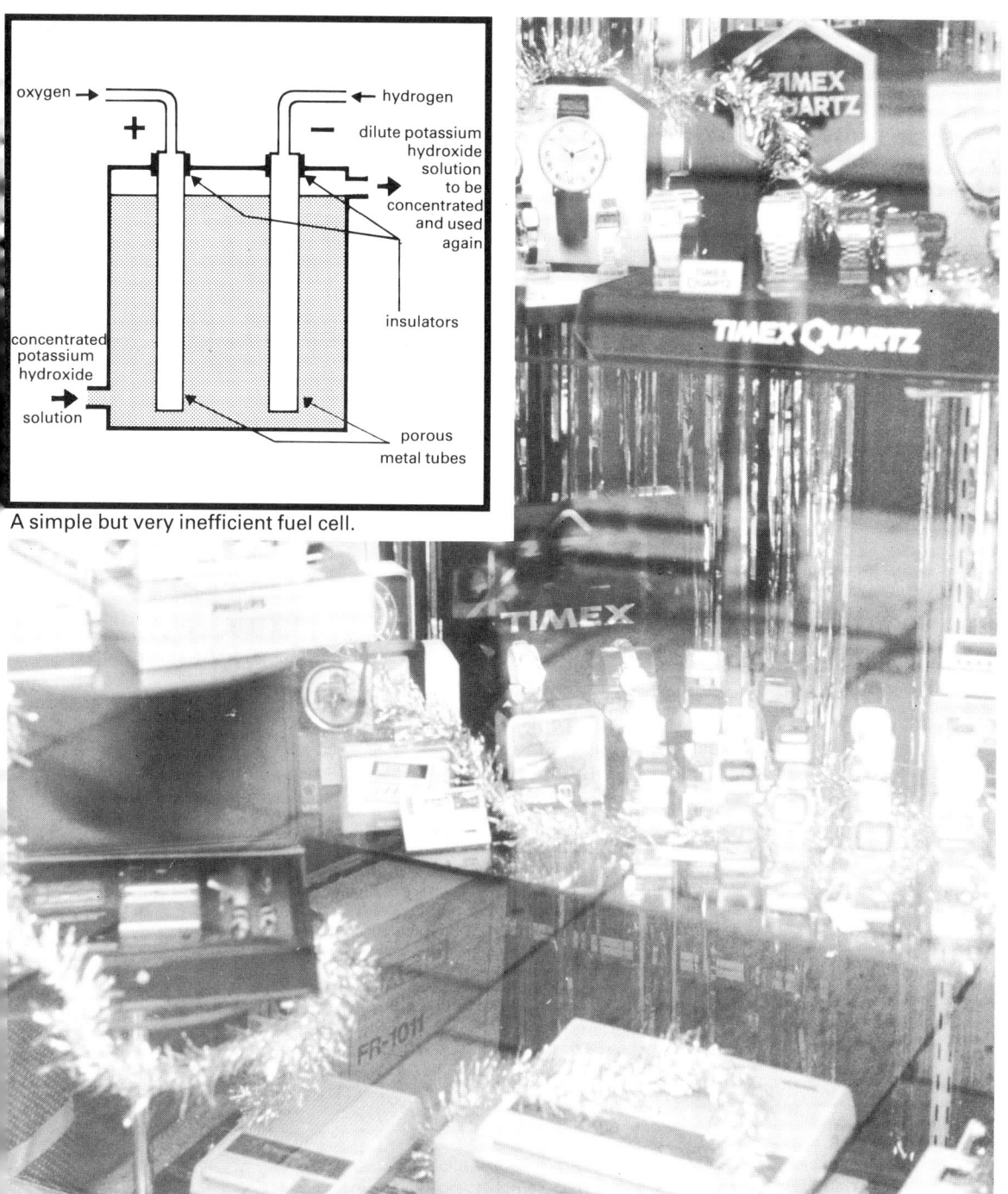

A simple but very inefficient fuel cell.

New Chemicals for Old

Just imagine you are living in the Stone Age, 30,000 years ago, what things would you miss? No sweets, no ice cream, no pop, no pens made of plastics, no penknife, no medicines – the list would be yards long if we continued. Stone Age man was very good at changing the shapes of things. He sewed skins together; he made tools of stone, and in general, used the materials around him. But, though he may have changed the shapes of his raw materials, he did not consciously alter their compositions to produce new substances. When he did learn to make new substances, such as wine from fruit juices, he had no idea of what was happening.

Fire was probably Man's first "tool" to make new substances. With it, he hardened wooden stakes by heating the ends till they were partially converted into charcoal; with it, he turned clay into pottery; with it, he learned to smelt metals from their ores. The burning of wood produces ash, smoke and heat. It is the heat that promotes the chemical reactions which change wood into charcoal, clay into pot, and green malachite into black copper oxide.

When a chemical reaction takes place, energy is involved. Sometimes energy is given out, as in burning. Such a reaction is said to be *exothermic* (exo = outside; thermic = heat). Sometimes

heat is taken in and the final products contain more energy within their structure than the starting chemicals. Such a reaction is said to be *endothermic* (endo = within).

Some reactions need a prod before they start. Take, for example, a match. It has to be struck before it will light. Air and petrol vapour in an internal combustion engine need a spark. The end of a cartridge must be hit by the firing pin of the gun before the gun will fire. In each case, a small amount of energy is needed to start the reaction, but once the reaction has been started, it produces enough energy to keep it going.

Heat isn't the only promoter of chemical reactions. Light can also promote chemical reactions, as, for example, when coloured dyes are faded by sunlight or when you take a photograph. Electricity can be used to break down solutions of metal salts, depositing the metal. This is what happens when objects are plated with silver or gold or chromium.

As a result of his chemical knowledge, Man is able to take crude oil and natural gas from the wells in the North Sea and turn them into plastics and drugs. Chemists can imitate Nature and make in factories copies of natural substances produced in the fields and forests of our land. Or they can make chemicals like nylon and polythene which do not exist naturally.

To produce synthetic materials for our use, chemists and technologists need Raw Materials, Knowledge, Factories, Staff to operate the factories, and Energy. Each of these factors is important, but the most important is Energy, because Energy is the coinage with which we pay for our modern "ships and shoes and sealing wax".

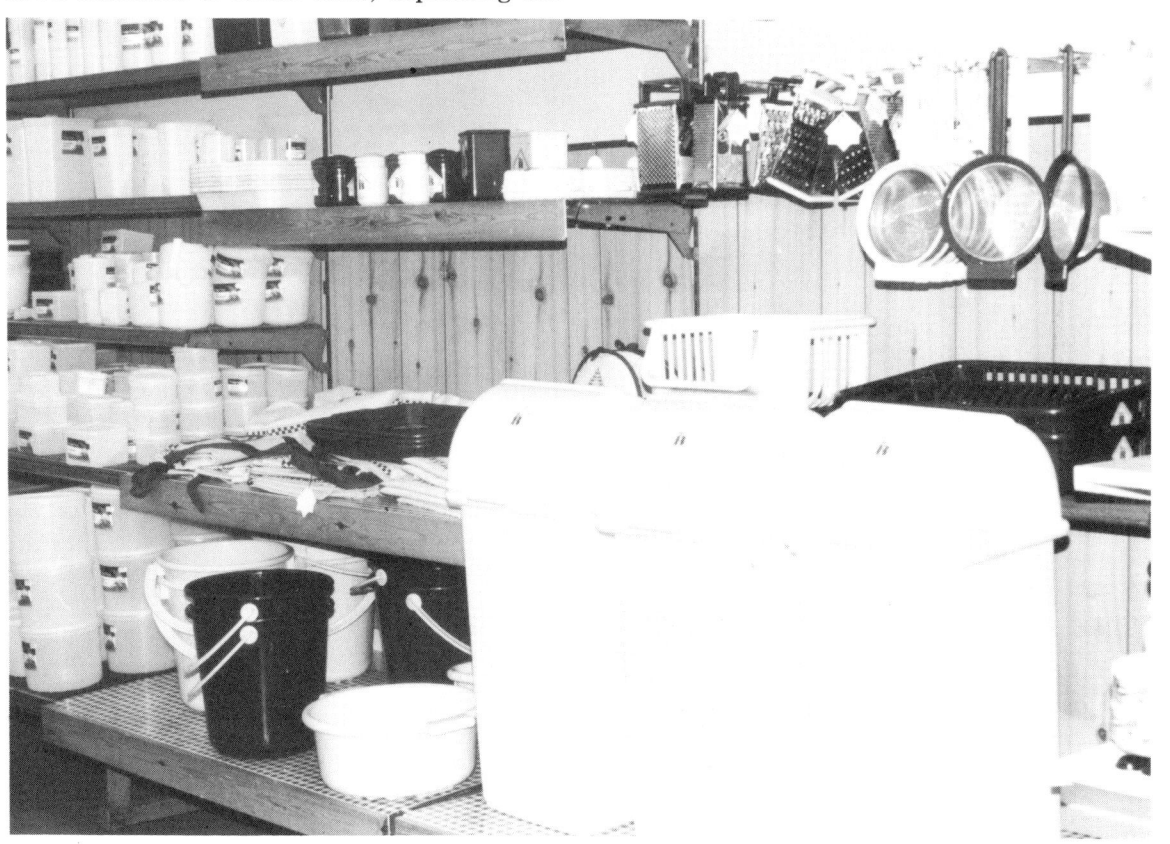

Solar Energy

Where did you go for holidays last year? Did you get sunny weather? Did you get sunburnt? Was it hot? Did it rain? Most of us when we go on holidays are sun lovers. We enjoy ourselves and think of the Sun as something that makes our holiday worthwhile. Yet, most of the energy we use everyday comes or has come from the Sun. If the Sun suddenly went out, like a switched-off electric light bulb, then it wouldn't be long before the Earth became a frozen, lifeless ball.

Of all the sunlight reaching the Earth, about 40% is reflected back into space by clouds and oceans. The rest lights and heats the Earth's surface. A world like Mars is colder and darker than Earth, because it is further away from the Sun. Mercury, the nearest planet to the Sun, has a surface temperature high enough to melt lead and zinc.

As a result of the heat from the Sun, together with oceans of water and an atmosphere, the

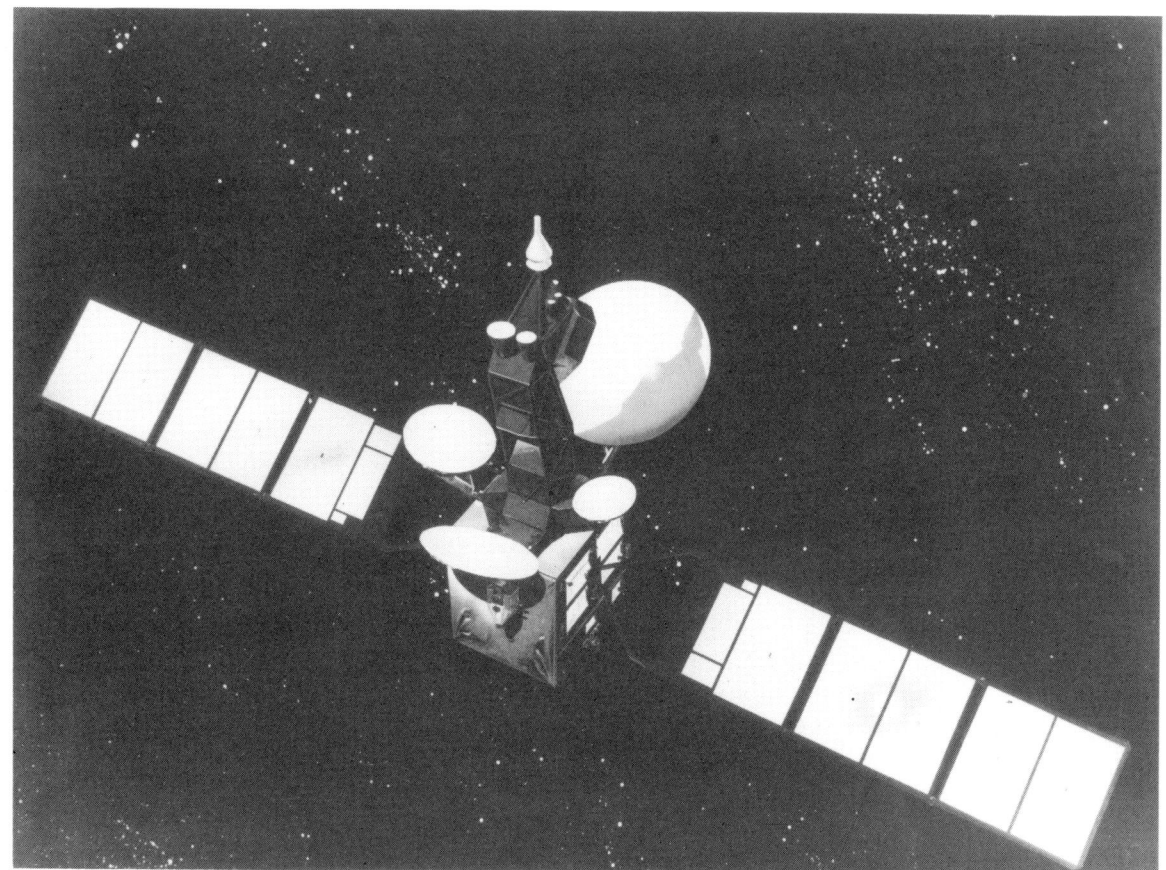

Earth has winds, rain and clouds – as can be seen on the satellite pictures shown on TV weather forecasts. The winds, rain and clouds are produced by something that has been called "The Weather Machine". Like all machines, it is powered by some form of energy, some form of "go". The Sun provides the "go".

We can use the "Weather Machine" to drive windmills, water-wheels and sailing ships. Recently, sunshine has been employed to provide domestic hot water from solar panels. Some space satellites, like Intelsat V, have big arrays of solar cells which convert sunlight directly into electricity.

All green plants make use of sunlight in photosynthesis (see page 12) to make sugars and starches from water and carbon dioxide. A by-product of photosynthesis is oxygen. All the oxygen in the air has been produced by green plants. We need oxygen to breathe. For food we eat green plants and their seeds and fruits. Our daily loaf is made of flour; the flour is ground wheat; the wheat depends on the Sun for energy to grow. For fuel we burn wood, coal and oil, all produced with the aid of sunlight, some many millions of years ago.

So, we rely upon the Sun for light and warmth, for plant life, for the air we breathe, for dawn and sunset, for the food we eat, and for the fuel that keeps us warm in winter and powers most of our machines, cars and aeroplanes. No wonder the ancient Egyptians worshipped the Sun as the God Ra.

Water Power

For much of Man's history he has had to rely upon muscle power. However, about 2000 years ago, a machine was invented which used another source of energy. This was the water wheel. It harnessed the energy of falling or flowing water and was first used to drive the millstones that ground the corn. Before the water wheel was introduced, it took two men or one donkey one hour to grind ten pounds of corn. A water wheel of about seven feet diameter could turn millstones that would grind about 400 pounds of corn in the same time. Water power reduced the number of freemen or slaves needed in the corn grinding trade and led to unemployment. But by the fourth century A.D. there was a shortage of manpower and so water mills flourished. Near Arles in France, sixteen water wheels were used to grind corn and produced three tons of flour an hour. This was sufficient to supply the needs of about 80,000 people.

The Roman Empire fell and then followed 500 dark years of invasions, wars and famines. Whole races migrated across Europe. The Vikings pillaged from the north. The Arabs struck from the south, crossing from North Africa and almost conquering Spain. By the eleventh century things had quietened down. Trade and the making of goods thrived. In England, south of the River Trent, the Domesday Book of 1086 listed 5,624 water mills. These were used not only for grinding corn but also for pumping water, sawing, hammering, grinding pigments and for many other industrial uses.

The water mill was the principal supplier of

A hydro-electric power station, as at Cruachan, shown in the photograph.

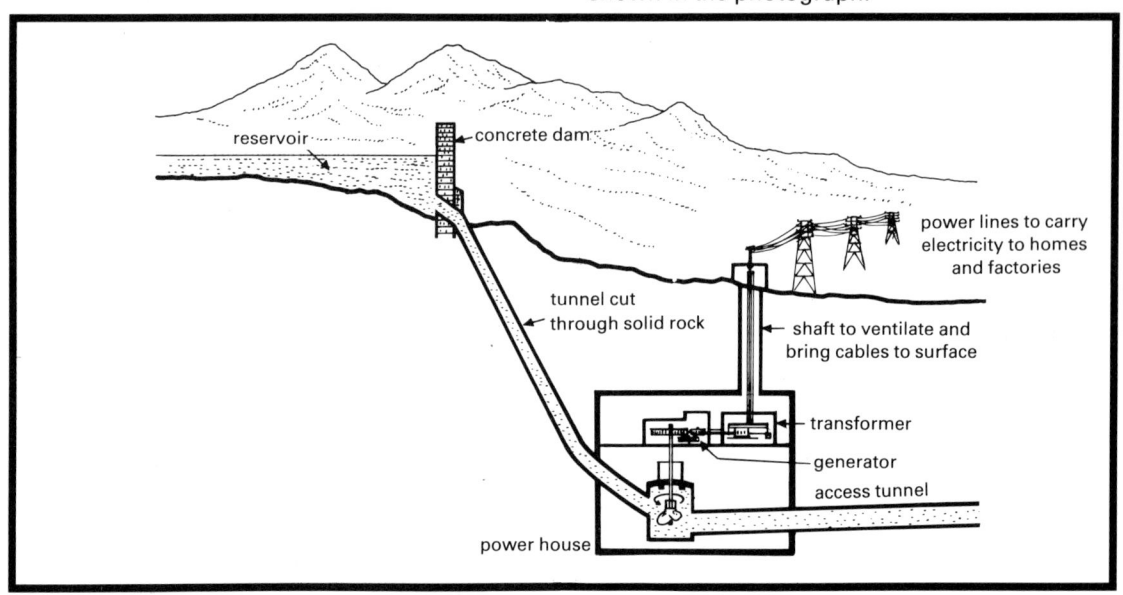

industrial energy until the coming of the steam engine. Then it slowly faded away. During the nineteenth century, when the steam engine became the power in the land, great inventions were made that improved the water wheel beyond recognition. Also, the dynamo, the principles of which were discovered by Faraday, was developed. This was a machine which converted Mechanical Energy into electricity. In 1886, work on the first great hydro-electric power station was started at Niagara Falls.

During the twentieth century, civil engineers have built many dams across river valleys. The impounded water has been used for irrigation and in hydro-electric schemes. The falling water is used to drive turbines and these turn electric generators. So, gravity helps provide cheap and convenient electricity without polluting the air with smoke, the land with ashes, and the lakes and rivers with acid. Steam power stations are economical and efficient, but they cannot be easily turned on or off. The demand for electricity varies according to the time of day, when people are travelling home from work, or when there is a special programme on TV. Hydro-electric power stations can be easily switched on to meet the demand. In the North of Scotland, 82% of the electricity generated is from hydro-electric power stations. But in England and Wales, where there are fewer opportunities for H.E.P., only three quarters of one per cent of the total production capacity is hydro-electric.

Windmills and Sailing Ships

We live in an ocean of restless air. Throughout Man's existence he has been used to the raging gales of winter and the breathless calms of summer. He has used these moving airs to power his ships and grind his corn. Even today, the jumbo-jets flying six or seven miles high are helped or hindered by streams of air travelling at speeds of up to 200 miles per hour. But the wind is a fickle mistress, blowing hot one day and cold the next; sometimes a gentle lamb, other times a roaring lion. None knew it better than the sailors of the days of sail. They knew of the Trade Winds, the Doldrums and the Roaring Forties. They put the winds to good use, just as the modern sailor uses them to power his racing yacht.

The harnessing of the winds on land took place probably only 800 years ago, whereas sailing ships have ploughed the seas for some 5000 years. Windmills, as we know them, came suddenly on the scene sometime in the twelfth century and quickly spread throughout Europe.

The windmill was the second of Man's practical mechanical machines to harness a source of energy other than muscle power, though there was an ingenious inventor, Hero of Alexandria, living in about 100 B.C., who made devices for opening temple doors by the expansion of air when heated. He even made a simple steam turbine, but his work was not pursued. His inventions were ingenious toys rather than practical machines.

At first, windmills were used to turn millstones which ground corn, but by the end of the fifteenth century they were used for all the same purposes as watermills. In the Low Countries they were used for pumping water. Just before the introduction of the steam engine, there were 8000 windmills keeping the low-lying land free of water.

The power output of a windmill with sails of 60 feet span was something like 10 horse power (7.5 KW) in a wind of 20 mph (10 metres per second). This is a very modest output and is no more than is produced by a modern small motor cycle engine. Although the energy is "free", it does depend on the weather. A high wind could cause the mill to "run away" and generate so much heat that it could easily catch fire. Fires were a hazard of the trade for the miller.

Although windmills were a benefit to Man, some people saw them as a curse. They believed that the mills would bring unemployment, as the watermill had done in Roman times. In 1767 a wind-powered saw mill was built at Limehouse, London. The following year it was wrecked by a mob.

Today, attempts are being made to generate electricity by modern windmills. Some experimental generators are being erected in the Orkneys, where strong winds blow most of the time. However, windmills employed to generate electricity for the electricity boards in South Wales and the USA have been less than successful, the main trouble being that the rotor blades frequently break. In today's world, wind power is not, so far, a viable alternative source of energy.

The Force that Drives the Sun

Q. What is the connection between the Sun, a nuclear power station and the Roman Baths in Bath?
A. They all have something to do with Nuclear Energy.

But what is Nuclear Energy? Do you remember that the Chemical Energy in a compound depends on how the atoms are put together in a molecule of the compound? The energy in an atom depends on the constituents of the atom, how many of them there are and how they are put together. As far as we are concerned, we can assume that the building bricks of atoms are protons, neutrons and electrons. The neutrons and protons form the nucleus of the atom, and this is normally surrounded by clouds of electrons. Very powerful forces bind the nucleus together. It is possible to obtain energy from the nuclei of certain atoms in two ways.

The first way of producing Atomic Energy is the method whereby the Sun gets its energy. It changes hydrogen into helium. Four atoms of hydrogen form one atom of helium. But, as four atoms of hydrogen have a greater mass than one atom of helium, the difference in mass is converted into energy. This is radiated from the surface of the Sun. Such a way of producing

energy is called the FUSION method, because atoms are fused, or joined together, to make larger atoms. The Sun is a gigantic nuclear furnace with a surface temperature of about 6000°C and an internal temperature of up to 13×10^{6}°C. Scientists are trying hard to produce energy by this means, but so far they have not been very successful.

The second method of producing nuclear energy is just the opposite of the fusion method. Large atoms are broken down into smaller atoms, with the liberation of energy. This is called the FISSION method. Natural fission radioactivity on Earth has been around ever since the Earth was formed, some four and a half thousand million years ago. In 1896, Antoine Henri Becquerel (1852-1908) discovered that the element uranium gave off rays that were capable of affecting photographic plates through layers of black paper. Work by Madame Curie and others established the existence of other radioactive substances that broke down without help into simpler substances, with the emission of invisible rays and particles. The Nuclear Age had begun.

It was soon found that natural radioactivity exists all around us, even in us. Seldom did it do us much harm. Radioactivity provided an answer to the question, "Why is the interior of the Earth hot?". Water from hot water springs, such as those at Bath, is warmed by the heat produced when atoms deep below the surface of the Earth break up into simpler substances.

Nuclear power stations use the heat produced in the radioactive fission of uranium, and elements formed from uranium, to heat water, producing steam. This steam, which is not radioactive, is used to drive turbines and electric generators.

Gravitational or Down-Hill Energy

It is hard work pedalling our bicycles up a hill. We have to use energy produced in our muscles to go up the hill. We have to work against the pull of the Earth, in the same way as we have to work lifting a weight from the floor on to the table. When we have got to the top of the hill, or when we have lifted the weight, we can get back some of the energy we have used. If we stay at the top of the hill or leave the weight on the table, it would be as if we had energy in the bank – stored energy. On our bicycle we can use the stored energy by freewheeling down the hill instead of pedalling.

On our bicycles we can freewheel down the hill and up the other side. But no matter how steep the hill which we freewheel down, we cannot freewheel up to a point higher than that from which we have started. We can use only the energy which we have stored lifting ourselves and the bike to the top of the hill. As we go down the hill, the stored energy is used to speed us along and to push the air away from us. Some energy is used up in the wheel bearings because, no matter how well we oil our bicycles, there is always some friction in the bearings. If there wasn't and we turned our bicycles upside down and spun a wheel, it would continue spinning forever. In the world of science you cannot get something for nothing. In fact, you cannot break even; in other words, if you pay 10p for energy you never receive 10p of useful energy. Some of it is wasted in various ways. Take our freewheeling ride down the hill and up the other side; energy is wasted in:

1. pushing the air to one side.
2. friction in the wheel bearings.
3. squeezing the rubber tyres as they come into contact with the road.

So, when we freewheel down the hill and up the other side, we end up at a lower point to that from which we started.

What about lifting a weight? Can we put it to some use? Sometime during the Middle Ages, possibly in the thirteenth century, Man discovered two uses for falling weights. In the first method, a falling weight was used to drive a clock. The first recorded mention of a clock in Britain was of one in Westminster in 1288. Weights have been used to power clocks up to modern times. Grandfather clocks, church clocks and many public clocks still use weights. The chiming and striking of Big Ben is powered by a weight of 1¼ tons.

The second use of descending weights was to power a war machine. This was called a TREBUCHET (treb-boo-shay). Look at the picture and see how it works.

Have you ever noticed how a rocket zooms up into the sky and then, in a graceful arc, falls back to earth? (Look at the picture on page 10.) Energy is needed to lift it high and then, when the store of Chemical Energy has been used up, the rocket falls. The path the rocket follows is known as a parabola. All rockets, whether they are fireworks or rocket-driven space craft, have to obey the same laws of Gravitation and Motion that were put forward by Sir Isaac Newton three hundred years ago. If you want to see his portrait, look on the back of a £1 note.

Other Sources of Energy

What other sources of energy are available? We have mentioned Oil, Coal, Wind, Water, Nuclear Power, Electro-Chemical Energy, Solar Energy, Electrical Energy and Gravitational Energy. Are there any other sources? If we look again at our list, we see that there are only 3 *primary* sources of energy:

Electro-Chemical: forces binding atoms together.
Nuclear: forces binding sub-atomic particles together.
Gravitational: forces binding the Earth, Moon, Sun and Stars together.

These give rise to *secondary* sources. For example, Electro-Chemical Energy can produce Heat and Electrical Energy; but neither Heat nor Electrical Energy are primary sources, because they are always made from another form of energy.

On pages 32-33, we dealt with Nuclear Energy and said that the internal heat of the Earth was due to natural fission radioactivity. Some parts of the Earth have volcanoes, hot springs and clouds of steam bursting out of the surface. So far we haven't found any use for volcanoes, but hot springs and steam jets have been put to good use. Iceland has plenty of hot springs and these are used to heat hot houses for growing plants. New Zealand and Italy make use of the natural supplies of steam to power electricity generating stations. (The photograph on page 37 shows the Wairakei Geo-Thermal Project in New Zealand.) In England, geologists have been looking for places where the Earth's internal hot water bottle is as close to the surface as it is in the City of Bath.

Another source of energy is in the ebb and flow of the waves of the oceans. Attempts are being made to tap tidal and wave power. Near St Malo in France a river estuary has been dammed and, as the tides rise and fall, they operate giant water turbines which are used to generate electricity.

The picture on the left shows how wave power can be used to cause floating boxes to nod up and down. This up-and-down motion drives an electrical generator. The nodding boxes are known as Salter's ducks and are named after Dr Salter of Edinburgh University. At present, they are still in the experimental stage.

It does not seem likely that there are any more unknown *primary* sources of energy waiting to be discovered. We have to make do with what we have. What need to be discovered are ways of tapping this energy.

Storing Mechanical Energy

Have you seen any films of Robin Hood or the film of Henry V at Agincourt? In them men use bows and arrows. Even today archery is a sport carried out at international level. What makes the arrow go? It is the bow straightening from its bent position. In a similar way, a piece of elastic can be stretched and then let go, as in a catapult slinging a stone. But the catapult or bow cannot do anything on its own. Energy has to be used to bend the bow or stretch the elastic. We supply the energy through the use of our muscles. It is stored, because the particles making up the bow or elastic have been pulled away from their usual position. When the bow or elastic is released, it flings the arrow or catapults the stone, and the particles take up their original positions.

Another way of storing energy is to use clockwork, as in a clockwork toy or wind-up alarm clock. Storing energy in elastic or a spring involves changing the shape of the object in which the energy is put; the stored energy is called Deformation Energy.

We can also store energy by lifting a weight, as in a grandfather clock or Big Ben, or by pumping water up into a storage tank. In these cases, the Mechanical Energy has been stored as Potential Energy.

If you turn a bicycle upside down and give the wheels a spin, they will carry on turning for a while, providing that the bearings are free and the brakes are not binding. If a dynamo is coupled to one of the wheels it will light a lamp until the wheel slows down. In giving the wheel a spin, you have put energy into turning the wheel. While it continues to turn, it retains or stores this energy and doles it out to overcome friction and air resistance and to drive the dynamo. The turning wheel acts as a store of energy.

In a similar way, steam and petrol engines use a fly-wheel to store energy so that the engine runs more smoothly. With steam engines, the fly-wheel is a large, heavy wheel. Some buses in Switzerland store the spare energy when descending a mountain, to speed up a high-speed fly-wheel. The stored energy is used to help the bus climb the next mountain.

The telephone exchange at Rochefort-en-Yvelines, west of Paris, has a fly-wheel spinning at 12,000 rpm, ready to power a dynamo and deliver 3 KW for 20 minutes, if the mains electricity fails.

To summarize, there are three ways of storing Mechanical Energy:

1. As Energy of Deformation, as in a spring or elastic or a bow.
2. As Gravitational Energy, as in a weight-driven clock.
3. In a rapidly rotating fly-wheel.

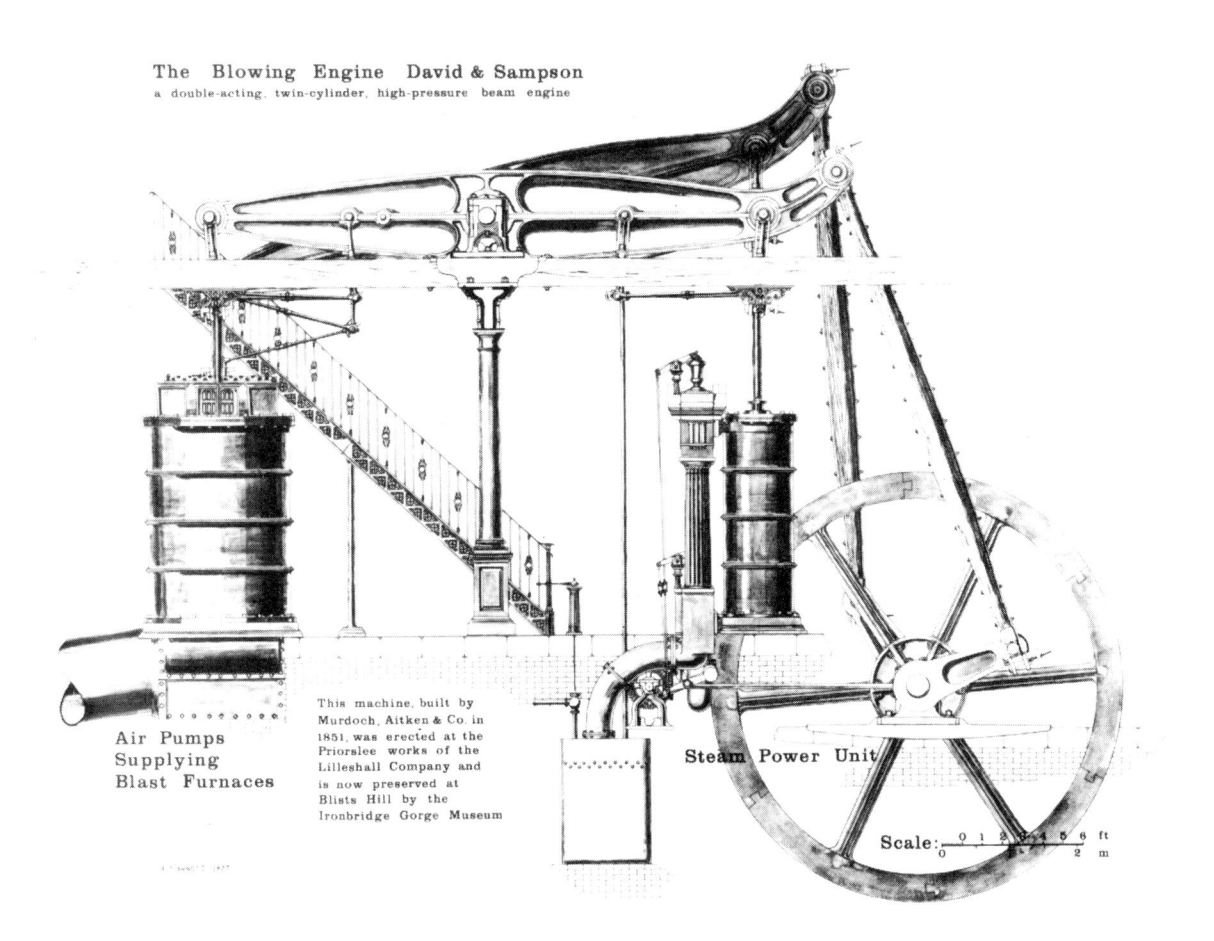

The Blowing Engine David & Sampson
a double-acting, twin-cylinder, high-pressure beam engine

Air Pumps Supplying Blast Furnaces

This machine, built by Murdoch, Aitken & Co. in 1851, was erected at the Priorslee works of the Lilleshall Company and is now preserved at Blists Hill by the Ironbridge Gorge Museum

Steam Power Unit

Storing Electrical Energy

Has your Mother or Father ever had trouble starting a car? What makes the starter motor go? Somewhere on a car is the starter battery. This provides Electrical Energy to turn the starter motor. When the engine is running, the dynamo or alternator charges the battery. The battery acts as a store of Electrical Energy.

Storing small quantities of electricity is easy. Storing large quantities of electricity is difficult. But it is easy to store water. Falling water can be used by a hydro-electric generator to produce electricity as the water flows from the upper reservoir to the lower reservoir. At night, when the demand for electricity falls, spare electricity from the Grid System that supplies electricity all over the country is used to work the hydro-electric generator in reverse. Water is pumped from the lower reservoir to the upper reservoir.

Pumped storage hydro-electric generator systems can be started at very short notice (about 2 minutes), whereas oil or coal-fired steam generators can take up to 8 hours to put into action. Pumped storage schemes are very useful because (1) they can be rapidly started up; and (2) the water can be used over and over again, so that the system does not need a large upper reservoir.

At present, much research is going on to find ways of storing in a battery enough electricity to power a motor car. Milk delivery vans are often powered by stored electricity. See if you can find out where the batteries are kept on the vans. It is cheap to run an electric van, but the batteries are heavy and the speed that can be achieved, while adequate for the stop-start run of a

Pumped hydro-electricity power station (a push-pull system).

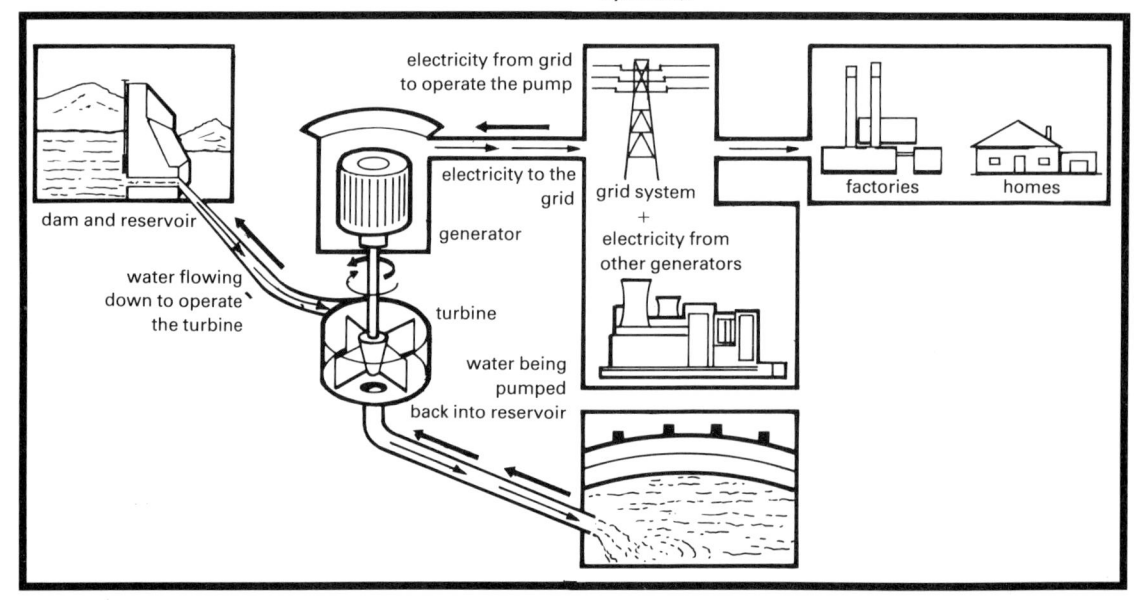

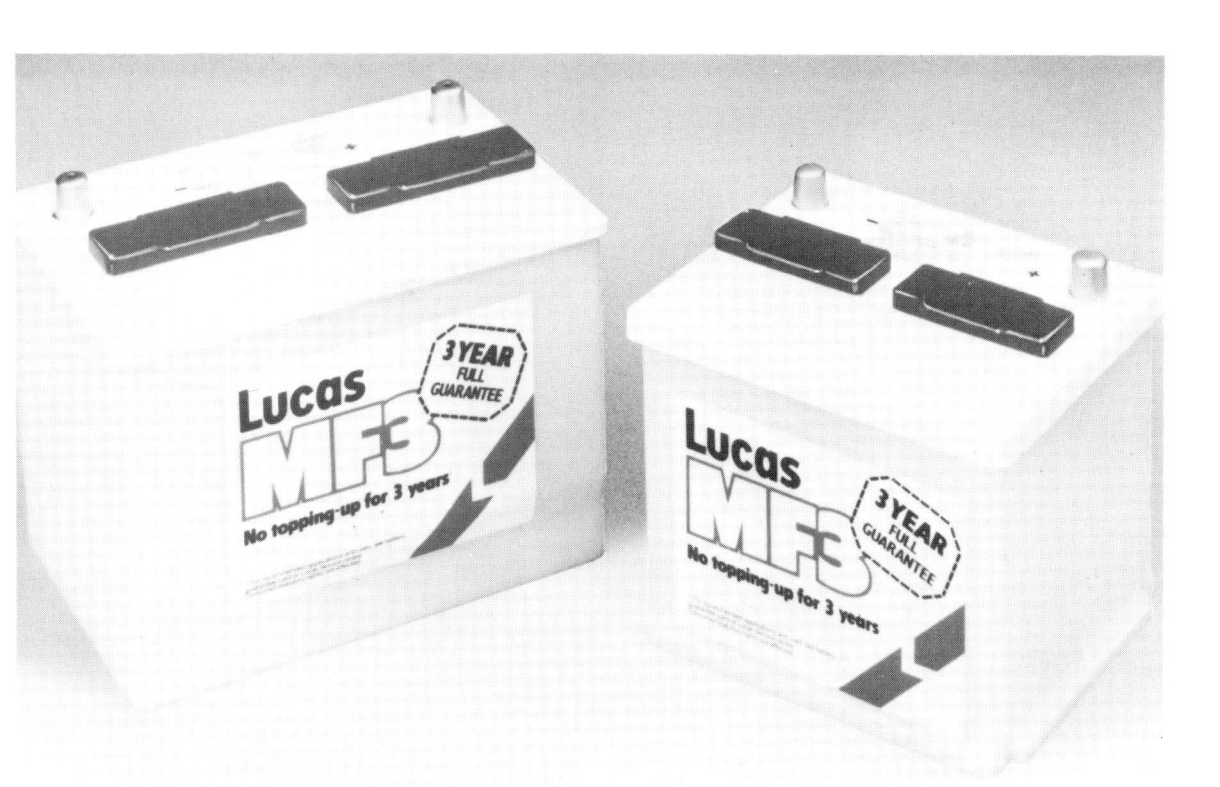

milkman, would not be enough for a motorist's everyday use. Some special cars using special batteries have been made, but they are very expensive to build and have a low top speed and limited range before the battery runs out.

Lucas, the makers of car electrical equipment, in collaboration with Reliant Motors, have designed a "Hybrid" car. It has an electrical motor and batteries for running around the town, and a small petrol engine to help when travelling long distances.

The Changing Faces of Energy

Energy is often like a second-hand car or motor cycle. It has passed through a number of hands before it reaches you. Look at sheep and cows comfortably grazing. They get their energy from their food – in this case, grass. Grass gets its energy from sunlight which has travelled 92 million miles from the Sun. In the Sun, the energy has been produced from the fusion of hydrogen into helium and the conversion of mass into energy.

Look at the moving electric train and trace the changes in energy. See how the ultimate source of energy is nuclear energy in the Sun.

Now consider a motor cycle. When the tank is being filled, stored fossil energy is being poured into the bike. Petroleum was formed millions of years ago as the result of the decay of plant and animal organisms. These organisms relied on the Sun for their energy. The energy was needed for living and growth and when the organisms died, the energy was fossilized in the oil. Today, much of our energy requirements are supplied by fossil fuel. In the cylinders of motor cycles petrol vapour burns in air. A chemical reaction takes place and heat is produced. This causes gases to expand and drive the engine. When slowing down, brakes are applied and the energy of movement is converted into heat.

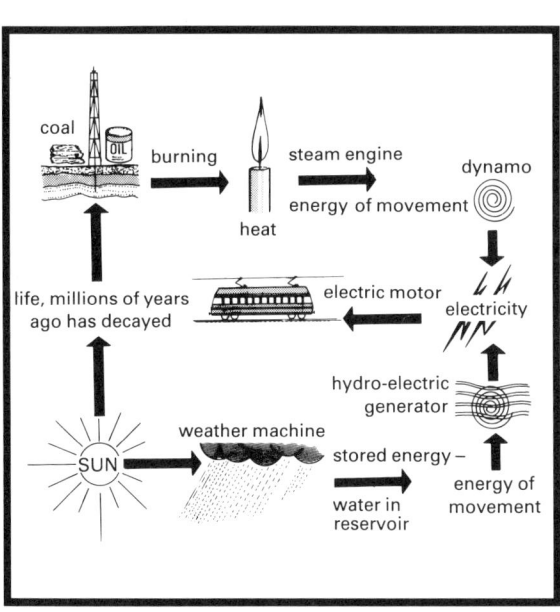

Do you notice that it is only when one form of energy is changed into another that we are able to make use of it? When this happens, there is always a fixed rate of exchange. Nature's store of energy cannot be destroyed; it can only be converted from one form into another. Every time this takes place, some of the energy is changed into useless heat. It is as if there is a Heat Tax on energy changes. Have you noticed that Heat always "runs down hill"? A hot water bottle will warm your cold feet; but your cold feet will not, on their own, make the hot water bottle hotter. In these last few sentences there have been put into everyday language two of the most important scientific laws of energy.

◁ How an electric train obtains its energy.

43

The "Go" in our Lives

What happens if our easy sources of energy, such as oil, coal and natural gas, run out? What effect would it have on you and me? Many of us would have no gas or electric cookers, no heating and no electric light in our homes. Our transport would be limited to muscle power. Many of the factories supplying us with ships and shoes and sealing wax would be shut. Look around you and see how energy has been used to make and bring to you all the bits and pieces of everyday life. Our civilization is founded upon energy. It is the one thing we need to keep us "going".

The United Kingdom is fortunate: it has large resources of coal and its coal mines produce over 100 million tons a year. It obtains natural gas, for cookers, and oil from gas and oil fields under the North Sea; and, at present, the country is independent of energy sources in the rest of the world – at least, almost, because it has to import quite an amount of food, and food is stored energy. All things considered, the UK is very fortunate, much more fortunate than a country such as Japan, which has to import nearly 80% of its energy and 99% of its oil. This means that Japan has to make things and sell them abroad so that it can buy energy.

Being fortunate does not mean that we can waste our energy resources. Energy in the form of coal, gas and oil is expensive. When we burn any of these to keep warm in the winter, we need to make sure that we do not waste heat by leaving windows and doors wide open. We must stop draughts and seek to insulate our buildings, so that heat is not lost through the walls and roofs. In our national newspapers there are advertisements seeking to persuade us to save energy by fitting double glazing or roof insulation or hot water tank jackets or cavity wall insulations. Look for these advertisements and see what they offer.

Once upon a time, Stone Age Man lived by the direct energy of the Sun. It supplied all his energy requirements, his food and his fuel; but, steadily, Man has increased his demands on energy. Nowadays, it is energy obtained mainly from sources other than directly from the Sun that makes possible all the items of our modern world. Without energy, we could not make the fertilizers that give us our crops of wheat; nor run the transport system that brings the wheat and bread and cakes to our shops.

Until the beginning of the Industrial Revolution in the eighteenth century, Man relied on Biologically Renewable Resources, like wood and charcoal and products of the Weather Machine, such as wind and water power. Then he started to use at an ever increasing rate irreplaceable fossil fuels, coal, then oil and later natural gas. At present, the civilization of the developed countries largely depends on these three sources of energy. We are using up Nature's energy storehouse and soon we will be driven to try to tame the fusion energy of the Sun itself. Energy sources such as Wind, Water, Wave, Tidal and Geothermal sources will help, but will not solve our problems. In the view of many scientists, our future depends on Nuclear Energy, because the sum total of all other sources, even if exploited to the full, will not give us enough energy to feed, clothe, house and transport our growing population. In large, expensive laboratories, using incredible machines and energies, a handful of scientists are working to save tomorrow. If they succeed, then it can be truly said that "Never was so much owed by so many to so few." We live in the most critical and most exciting period of Man's history.

£109.20 OFF
Full price £408.20
Offer price £299
£80.87 OFF
Full price £379.87
Offer price £299
£105.88 OFF
Offer price £319
£20 TRADE 'IN' ON THIS APPLIANCE
Cooker of the Month
Plus 12 months Interest Free Credit
Plus 12 months Interest Free Credit

Glossary

atoms: the smallest portions of an element that can exist on their own and take part in a chemical reaction.

biologically renewable material: material that, because it is derived from living material, can be replaced by growth.

dynamo: machine that will convert Mechanical Energy into Electrical Energy, also called an *Electrical Generator.*

endothermic reaction: a reaction in which heat is taken in.

electrons: the smallest particles of negative electricity: they are elementary particles found in atoms.

exothermic reaction: a reaction in which heat is given out.

fossil fuel: any fuel derived from living organisms that has been preserved by natural means, on or under the Earth's surface. Eg, oil, peat, coal.

fuel cell: a device for producing electricity by the oxidation of a fuel: Chemical Energy is converted directly into electricity.

galaxy: a cluster of stars. Our own galaxy, to which the Sun belongs, is called the Milky Way. There are millions of other galaxies.

horse power: the British unit of power; it is work done at the rate of 550 foot-pounds per second. (A foot-pound is the energy required to lift a pound weight through one foot in a second.) 1 HP = 745.7 Watts.

ignition coil: used in some internal combustion engines to provide high voltage (30,000V) electricity. *See* spark plugs.

malachite: a green mineral containing basic copper carbonate.

mass: a measure of how much stuff is in a body. Mass is independent of the pull of the Earth.

methanol: a chemical compound having the formula CH_3OH. It is the first member of a series of alcohols. The next member is Ethanol which is found in wines, spirits and ales. Methanol is very poisonous.

neutron: an elementary particle, of atomic weight 1, found in all atomic nuclei except hydrogen.

nucleus of an atom: the positively charged centre of an atom. Nearly all the mass of an atom is concentrated in its nucleus, but the volume of the nucleus is only a very small fraction of the total volume of the atom.

precipitate: the insoluble material formed when two reacting solutions are mixed.

propeller: a fan-like device rotated by an engine to speed up the flow of water or air.

proton: an elementary particle found in all atomic nuclei. It has a positive electrical charge, equal but opposite to that of an electron. It has a mass almost equal to the mass of a single hydrogen atom.

radioactivity: the property possessed by some atomic nuclei in which the nucleus naturally breaks up into smaller particles. The process is accompanied by rays of different kinds.

spark plugs: used in some internal combustion engines. High voltage from the ignition coil sparks between two electrodes, causing the air-petrol vapour to explode. Diesel engines do not require ignition coils or spark plugs.

sub-atomic particles: the particles that form the structure of an atom. Electrons, neutrons and protons are examples of sub-atomic particles; there are a number of others.

train of gears: a number of gear wheels driving one another so that power can be transferred from one place to another.

trebuchet: war machine invented in the twelfth or thirteenth century. Powered by Gravitational Energy, it flung stones at the enemy.

turbines: devices powered by steam, burning gases or running water; they are used to drive aeroplanes, ships and electrical generators.

vacuum: strictly speaking, this means empty space, but a perfect vacuum is unobtainable. Vacuum is generally taken to mean a space containing a gas at a very low pressure.

vitamins: a number of organic substances present in food that are essential for a healthy life.

Weather Machine: The weather and climate systems are powered by the Sun, which provides heat energy to drive the winds and storms.

Book List

Calder, R., et al., *Energy,* Time-Life, 1969
Chant, C., *Aviation, an Illustrated History,* Orbis Publishing Ltd, 1980
Dubos, R., et al., *Energy,* Time-Life, 1969
Jagger, C., *The World's Great Clocks and Watches,* W.H. Smith, 1981
Mabey, R., *The Pollution Handbook,* Penguin Books Ltd, 1974
Stone, R., Dennien, R., *Energy,* Longman Group Ltd, 1974
Strandh, S., *Machines, an Illustrated History,* A.B. Nordbok, Gothenburg, Sweden, 1982
Vince, J., *Discovering Watermills,* Shire Publications, 1980

Booklets on the production of electricity by Wind, Water, Nuclear, Oil and Coal powered generators are available from:

>Central Electricity Generating Board,
>Sudbury House, 15 Newgate Street, London EC1.

>North of Scotland Hydro-Electric Board,
>16 Rothesay Terrace, Edinburgh 3.

>South of Scotland Electricity Board,
>Spean Street, Glasgow.

>U.K. Atomic Energy Authority, (Information Services Branch),
>11 Charles II Street, London SW1Y 4QP.

Index

archery, 38
Arles, watermill 28
Aten, 5

Bath, hot waters, 32, 33, 36
batteries, 22
Becquerel, Antoine Henri, 33
Big Ben, 10, 11, 35, 38
Boulton, Matthew, 16

Calder Hall, 10, 11
clocks, 35
Concorde, 20, 21
Cruachan hydro-electric
   scheme 28, 29

diesel engines, 19
Diesel, Rudolf, 19
Domesday book and watermills, 28
dry cells, 22
dynamo, 29

electric cars, 40, 41
electrolyte, 22
endothermic reactions, 25
energy,
   biologically renewable
     resources, 44
   changes, 42, 43
   chemical, 10, 12, 22, 23, 35
   deformation, 38-39
   electrical, 8, 10
   electro-magnetic
     radiation, 8, 9, 10
   geothermal, 32, 33, 36, 37
   gravitational, 10, 34, 35, 38-39
   heat, 12, 22, 24, 36, 42, 43, 44
   kinetic, 8
   mechanical, 8, 10, 22, 29, 38-39
   nuclear, 10, 32, 33, 36
   potential, 10, 38, 39, 40, 41
   primary sources, 36
   resources, Japan, 44
     Stone Age man, 44
     U.K., 44
   solar, 26, 27, 44
   storing, mechanical, 38, 39
     electrical, 40, 41

   tidal, 37
exothermic reactions, 24
explosives, 14, 15

fire, as a tool, 24
fly-wheel, 38, 39
fuel cell, 22, 23
fuels, 22

geothermal power, 36, 37
grinding corn, 28

Hero of Alexandria, 30
hydro-electric power, 28, 29

Industrial Revolution and steam
   engine, 16
Intelsat V, 27
internal combustion engine, 18, 19
Ironbridge C.E.G.B. steam
   turbine, 20, 21

jet engine, 20

lithium, 22
Lucas, batteries and electric
   cars, 41

manganese dioxide, 22
Mars, 26
matches, 14
Maxwell, James Clerk, 6
mercury, 26
methanol, 22
Mitchell, R.J., 19
Moon, 10
muscle power, 12, 13
muscle power, human output, 12

Newcomen, Thomas, 16
Newcomen engine, efficiency
   and how it worked, 16, 17
Newton, Sir Isaac, 10, 35
Niagara Falls and hydro-
   electricity, 29
nuclear energy,
   discovery of radioactivity, 33
   natural, 32, 33

   fission method, 33
   fusion method, 33

petroleum, formation, 42
photosynthesis, 27
promoters of chemical
   reactions, 25
pumped hydro-electric power
   station, 40

Ra, 27
radio waves, 9
Rochefort-en-Yvelines, 39
rockets, 14, 15
Rolls Royce engines, 19

St Malo, tidal energy, 37
Salter's Ducks, 36, 37
satellite pictures, 26, 27
Schneider Trophy
   aeroplanes, 19
slave power, 13
space satellite, 27
Spitfire, 19
steam engines, 16, 17
steam engines and Industrial
   Revolution, 16
Swiss buses and fly-wheels, 39
synthetic material, 25

trebuchet, 35
turbines, gas, 20, 21
   steam, 20

velocity of electro-magnetic
   radiation, 8

Wairakei Geo-thermal
   Project, 36, 37
watermills, 28, 30, 31
watermills, power output, 31
Watt, James, 16
wave-lengths of electro-magnetic
   radiation, 9
weather machine, 27
windmill, 20
windmill, Orkneys, 31
   sawmill, Limehouse, 31

zinc, 22